HRH Prince John Charles Wright

HRH Prince Joe Duncan Wright

Metallurgy, mining and steel
manufacturing in Rhodesia

1st Published in 1982

2nd Publication in 2002

Table of Contents

Foreword

For as long as we can remember, our family has for generations had an interest in metallurgy including processes such as mining, ore beneficiation, smelting, and refining to obtain pure metals or metal alloys. Countless hours have been spent studying metals, their properties, and their production and use. Understanding the behaviour of metallic elements and alloys, their extraction from ores, and the various processes used to modify their properties for different applications. Metallurgical knowledge and expertise are essential for industries such as manufacturing, construction, automotive, aerospace, energy, and many others that rely on metals and alloys. By understanding the principles of metallurgy, engineers and scientists can develop new materials, improve manufacturing processes, enhance material performance, and ensure the reliability and safety of metallic products.

Sustainable extractive metallurgy involves minimizing the environmental impact of mining and extraction activities. This includes implementing technologies and practices that reduce water usage, waste generation, and emissions of harmful substances. Recycling and reuse of process water, proper waste management, and adherence to environmental regulations are important aspects of effective extractive metallurgy. Sustainable extractive metallurgy involves the efficient and sustainable extraction of metals from their ores or raw materials. Choosing the most suitable extraction process is crucial for efficient metallurgical operations. The process selection should consider factors such as ore grade, mineralogy, energy consumption, environmental impact, and economic feasibility. Continual process optimization helps maximize metal recovery, minimize waste generation, and reduce energy and water consumption. It involves optimizing parameters like particle size distribution, pH, temperature, reagent dosage, residence time, and equipment configuration. Advanced modelling and simulation tools can be utilized to analyze and optimize the process performance. Energy consumption is a significant factor in extractive metallurgy. Implementing energy-efficient technologies and processes helps reduce operational costs and minimize the environmental footprint. This can include using advanced grinding technologies, optimizing smelting and refining processes, utilizing renewable energy sources, and recovering waste heat.

Mining in Rhodesia, now known as Zimbabwe, has required a combination of factors and practices to ensure optimal extraction and utilization of mineral resources. There were some key aspects that contribute to effective mining in Rhodesia. Detailed geological surveys and exploration were essential to identify and evaluate mineral deposits. Comprehensive knowledge of the geology and mineral potential of the region helped guide mining operations and optimize resource extraction. Adequate infrastructure, including transportation networks, power supply, and water resources, have proved crucial for efficient mining operations. Investment in infrastructure facilitates the movement of equipment, supplies, and personnel, as well as the transportation of mined materials.

Adequate infrastructure is crucial for effective mining operations. This includes transportation networks (roads, railways, and ports) for moving equipment and extracted minerals, access to reliable power supply, and water resources for mining and processing activities. Investments in infrastructure support the overall efficiency and productivity of mining operations. Maximising the recovery of valuable minerals and optimizing the mineral processing and refining techniques are crucial for effective mining. Employing efficient extraction, concentration, and refining processes can increase yields and reduce waste, maximizing the value obtained from the extracted minerals.

Adopting modern mining techniques and technologies improves operational efficiency and safety. Mechanised and automated equipment, remote sensing, and advanced geological modelling techniques enhanced exploration, extraction, and processing capabilities. Ensuring the safety and well-being of workers has been of utmost importance. Compliance with safety regulations, proper training, and the provision of personal protective equipment has been crucial. Additionally, implementing environmentally responsible practices, such as proper waste management and reclamation plans, minimises the ecological impact of mining operations.

A clear and stable regulatory framework provides a predictable environment for mining operations. Transparent licensing procedures, legal protections, and consistent enforcement of regulations contribute to effective mining practices. Good governance and anti-corruption measures are also important for maintaining a favourable investment climate. A skilled and knowledgeable workforce has been essential for effective mining in Rhodesia Engaging with local communities, government authorities, and other stakeholders fosters positive relationships and ensures that mining benefits are shared equitably. Consultation with affected communities, addressing social and environmental concerns, and contributing to local development initiatives enhance the social license to operate. Training programs and continuous education help develop competent professionals who can operate mining equipment, handle technical aspects of mining processes, and implement safety protocols.

Ensuring a reliable and consistent supply of raw materials, such as iron ore, coal, and fluxes, has been crucial for steel manufacturing. Developing partnerships or securing long-term contracts with reliable suppliers has helped maintain a steady supply chain and minimizes disruptions in production. Implementing efficient production processes is essential for maximizing productivity and reducing costs. This includes optimizing the operation of blast furnaces, steelmaking converters, and rolling mills. Employing advanced technologies, automation, and process control systems can enhance operational efficiency and product quality. Having a skilled and knowledgeable workforce is critical for effective steel manufacturing. Providing comprehensive training programs and ongoing professional development can improve the competency of employees and enable them to operate equipment and carry out tasks efficiently. Investing in workforce development ensures a high level of expertise and contributes to the success of steel manufacturing operations.

Steel manufacturing is an energy-intensive process. Ensuring a reliable and cost-effective energy supply, such as electricity and fuel, is important for maintaining competitiveness. Implementing energy-saving measures, optimizing energy consumption, and exploring renewable energy sources can help manage energy costs and reduce the environmental impact. Maintaining consistent product quality and meeting customer specifications are essential for the success of steel manufacturing. Implementing stringent quality control measures throughout the production process ensures that the final steel products meet the required standards. Additionally, investing in research and development for product improvement and diversification can open new market opportunities and enhance competitiveness.

Adequate infrastructure, including transportation networks, storage facilities, and access to ports, is critical for the efficient movement of raw materials and finished steel products. Investing in infrastructure development and optimizing logistics operations can streamline the supply chain and reduce transportation costs. Steel manufacturing should prioritize environmental and social responsibility. Adhering to environmental regulations, implementing sustainable practices, and minimizing the environmental impact of manufacturing operations are crucial. Engaging with local communities, promoting safe working conditions, and supporting social development initiatives contribute to a positive reputation and long-term sustainability. These factors, when properly addressed and managed, contributed to effective mining practices in Rhodesia, as well as in any mining region. However, it's important to note that the steel manufacturing and mining industry is dynamic and constantly evolving, and ongoing monitoring and adaptation to changing circumstances are necessary for sustainable and effective mining practices.– HRH Prince John Charles Wright and HRH Prince Joe Duncan Wright

Mining in Rhodesia

The first step in extractive metallurgy is the extraction of ores from the Earth's crust. Ores are naturally occurring rocks or minerals that contain economically valuable metals. Mining methods depend on factors such as the depth and location of the ore deposit, and can include surface mining (open-pit or strip mining) or underground mining. Extractive metallurgy plays a vital role in the production of metals for various industries, including automotive, aerospace, construction, electronics, and more. It requires a deep understanding of the physical and chemical properties of metals and ores, as well as expertise in mining, mineral processing, smelting, refining, and alloying techniques. Continuous research and development in extractive metallurgy are important to improve efficiency, reduce environmental impacts, and discover new methods for extracting and refining metals.

Mining was indeed an important sector in Rhodesia, with the country being a significant producer of various minerals. Mining was an important sector in Rhodesia, and the country was a significant producer of various minerals including gold, chrome, asbestos, and coal. Rhodesia had rich mineral resources, and mining played a vital role in its economy. Some of the key minerals produced in Rhodesia included gold, chrome, asbestos, chrome and other minerals. These minerals, along with others such as copper, nickel, and tin, contributed to Rhodesia's mining sector and played a crucial role in its economic development. Mining played a vital role in the economy of Rhodesia, now known as Zimbabwe, during its existence. The country was rich in various minerals, and mining activities contributed significantly to its economic development.

Rhodesia had substantial gold deposits, and gold mining was a major industry. It attracted significant investment and contributed to foreign exchange earnings. Rhodesia was known for its rich chrome ore deposits. Chrome mining and production were important industries, with the country being a major global producer of chromium. Rhodesia had significant asbestos deposits, particularly in the Shabani and Mashaba areas. Asbestos mining and production were major industries, with the country being one of the leading global producers of asbestos.

Rhodesia had coal deposits in areas such as Hwange, Wankie, and Wankie Colliery, which played a role in meeting domestic energy needs and supporting industrial activities. In addition to these minerals, Rhodesia also had deposits of other resources such as copper, nickel, tin, limestone, and more. Mining activities in these sectors contributed to the overall economic growth of the country. The mining industry in Rhodesia attracted both local and foreign investment and provided employment opportunities to a significant number of workers. It also supported infrastructure development, such as roads, railways, and power supply, in the mining regions. However, it's important to note that the mining industry in Zimbabwe, formerly Rhodesia, has experienced various challenges and changes over time due to factors such as political instability, economic shifts, and changes in mining policies.

Gold

Gold mining in Rhodesia can be traced back to ancient times, with evidence of gold mining activities by indigenous people. However, it was during the colonial period that large-scale gold mining operations began to take shape. The discovery of gold in the 19th century attracted a wave of prospectors and miners to Rhodesia. The industry experienced rapid growth, and mining companies were established to exploit the gold deposits.

One of the most famous gold mining areas in Rhodesia was the Mashonaland region, particularly the areas around the capital city of Salisbury (now Harare). Other significant gold mining areas included Matabeleland and the Midlands. Rhodesia had substantial gold deposits, and gold mining played a crucial role in the country's economy. It attracted significant investment and contributed to foreign exchange earnings. Rhodesia had substantial gold deposits, and mining played a crucial role in the country's economy. Gold mining operations were carried out in various regions, including the areas around the cities of Bulawayo, Gwanda, and Kadoma. The gold industry contributed significantly to foreign exchange earnings and employment opportunities.

Gold mining played a significant role in Rhodesia's economy during its existence. The country had substantial gold deposits, attracting considerable attention from local and international mining companies. Rhodesia had various gold mining regions, including the famous goldfields of the Mazowe, Shamva, and Bulawayo districts. These areas were known for their rich gold deposits and attracted many prospectors and miners. Gold mining operations in Rhodesia ranged from small-scale artisanal mining to large-scale industrial operations. Companies such as Rio Tinto, Anglo American Corporation, and Lonrho had a presence in gold mining and exploration in the country.

The revenue generated from gold mining contributed to Rhodesia's economy, providing employment opportunities and foreign exchange earnings. Gold was a valuable commodity, and its production played a crucial role in the country's overall economic development. However, it's important to note that the political and economic shifts in Rhodesia, including the country's transition to Zimbabwe in 1980, had an impact on the gold mining sector. Changes in government policies, land reforms, and other factors influenced the industry's dynamics over time.

Gold mining was a significant industry in Rhodesia, which is now known as Zimbabwe, during its colonial era. Rhodesia had rich gold deposits, and gold mining played a crucial role in the country's economy and development. The gold mining industry in Rhodesia involved both large-scale industrial operations and small-scale artisanal mining. Large mining companies, such as the Rhodesian Mining and Land Company, played a dominant role in gold extraction, employing advanced technologies and machinery. These companies operated deep-level mines and employed a substantial number of workers.

Small-scale miners, known as "artisanal miners," also played a significant role in gold mining. These were individuals or small groups who engaged in mining activities with simpler tools and methods, often working in informal or illegal operations.

Gold mining contributed to Rhodesia's economy by generating revenue through exports and attracting foreign investment. It provided employment opportunities, infrastructure development, and technological advancements. The industry also had spin-off effects on related sectors, such as mining equipment manufacturing, transportation, and financial services.

Gold mining played a significant role in Rhodesia's economy during its existence. The country had substantial gold deposits, and gold mining was an important industry. Rhodesia had several gold mining regions, including the Kwekwe, Kadoma, and Gwanda areas. These regions were known for their rich gold reserves and attracted both local and foreign investment in mining operations. Gold mining in Rhodesia involved both large-scale commercial mining and small-scale artisanal mining. Large mining companies, such as Lonrho and Anglo American Corporation, operated in the country and employed a significant number of workers.

The gold mining industry in Rhodesia contributed to the country's economic growth, provided employment opportunities, and generated foreign exchange earnings through gold exports. It played a vital role in supporting infrastructure development, such as roads, railways, and power supply, in the mining regions.

However, political and economic challenges, including the transition to Zimbabwe and changes in mining policies, had an impact on the gold mining industry. Factors such as land redistribution, changes in ownership structures, and fluctuating gold prices influenced the industry's operations and investment climate.

After the country gained independence from British colonial rule in 1980 and became Zimbabwe, the gold mining industry underwent significant changes. The government introduced new mining policies and regulations, and the sector experienced shifts in ownership and management. After Rhodesia gained independence from British colonial rule in 1980 and became Zimbabwe, the gold mining industry went through various changes and challenges. Economic and political factors, such as land reforms and changes in mining regulations, had an impact on the industry's dynamics.

Today, gold mining continues to be a significant industry in Zimbabwe, contributing to the country's economy and attracting investment. The sector includes both large-scale mining operations and artisanal miners.

Chrome

Rhodesia was known for its rich chrome ore deposits. Rhodesia, which is now known as Zimbabwe, had significant chrome mining operations. Chrome mining was a prominent industry in the country and contributed to its economy and export revenues.

Rhodesia had rich chrome ore deposits, particularly in the Great Dyke region. The Great Dyke is a geological feature that stretches across Zimbabwe and contains substantial reserves of chrome and other minerals. The chrome deposits in Rhodesia were of high quality and attracted interest from both local and foreign mining companies.

Chrome mining in Rhodesia involved both large-scale commercial operations and smaller-scale mining activities. Companies such as Rhodesian Mining and Anglo American Corporation were involved in chrome mining and exploration. The mining operations employed a significant workforce and contributed to the overall employment and economic development of the country.

The chrome extracted from Rhodesia was processed into ferrochrome, a vital alloy used in stainless steel production. Chrome ore was also exported to other countries, contributing to foreign exchange earnings for the country.

However, similar to other sectors in Rhodesia, the chrome mining industry faced challenges due to political and economic changes. After the country gained independence in 1980 and became Zimbabwe, the government implemented new mining policies and regulations, which had an impact on the industry's operations and investment climate.

Chrome mining and production were important industries, with the country being a major global producer of chromium. Rhodesia also had extensive reserves of chromium ore, which is used in the production of stainless steel and other alloys. Chrome mining operations were primarily concentrated in the Midlands province, particularly around the towns of Kwekwe and Shurugwi.

Asbestos

Asbestos mining was a significant industry in Rhodesia, now known as Zimbabwe, during its existence. Rhodesia had substantial asbestos deposits and was one of the leading global producers of asbestos.

The main areas of asbestos mining in Rhodesia were the Shabani and Mashaba regions, which had large asbestos mines. Asbestos, a naturally occurring mineral known for its heat resistance and durability, was extensively mined in these areas. The asbestos mining industry in Rhodesia contributed to the country's economy, employment, and export revenues. The asbestos fibres were used in various industries, including construction, insulation, and manufacturing, due to their desirable properties.

However, it's important to note that the health risks associated with asbestos exposure became widely recognized over time. Asbestos fibres, when inhaled, can cause serious health issues such as lung diseases and cancer. As awareness of these health risks grew, the use and mining of asbestos declined worldwide.

After the country gained independence from British colonial rule in 1980 and became Zimbabwe, the asbestos mining industry faced significant challenges. The government implemented stricter regulations and safety measures to protect workers and the environment from asbestos-related hazards. Over time, the asbestos mining industry in Zimbabwe has decreased, and efforts have been made to transition away from asbestos mining and promote safer alternatives.

Rhodesia had significant asbestos deposits, particularly in the Shabani and Mashaba areas. Asbestos mining and production were major industries, with the country being one of the leading global producers of asbestos at that time. Rhodesia was known for its vast deposits of asbestos, particularly the Chrysotile variety. Asbestos mining took place in several regions, including Shabani (now Zvishavane) and Mashaba. However, the health risks associated with asbestos led to the decline of the industry in later years.

Asbestos mining poses significant risks to human health and the environment. The health hazards associated with asbestos exposure are well-documented, and precautions must be taken to minimize the risks involved.

The primary risk associated with asbestos mining is the inhalation of asbestos fibres. Prolonged or heavy exposure to asbestos fibres can lead to serious health conditions, including lung diseases such as asbestosis, lung cancer, and mesothelioma, a rare and aggressive form of cancer.

Miners working in asbestos mines are particularly vulnerable to exposure. The mining process itself, which involves drilling, blasting, and extracting the asbestos-containing rock, releases asbestos fibres into the air. Additionally, handling and processing asbestos ore can further release fibres.

Controlling and managing the risks associated with asbestos mining requires strict adherence to safety protocols and regulations. This includes providing miners with appropriate personal protective equipment (PPE), implementing proper ventilation systems, and conducting regular monitoring of airborne asbestos fibre levels. It is also crucial to provide comprehensive training to workers on asbestos hazards, safe work practices, and proper decontamination procedures.

Environmental risks are also associated with asbestos mining. Asbestos fibres released into the environment during mining operations can contaminate air, water, and soil. These fibres can persist in the environment for a long time and pose a risk to nearby communities and ecosystems.

Due to the well-documented health risks, many countries have implemented strict regulations or outright bans on asbestos mining and the use of asbestos-containing products. These measures aim to protect human health and prevent further exposure to asbestos fibres.

Coal

Coal mining was another significant sector in Rhodesia. Coal mining was an important industry in Rhodesia, contributing to the country's energy needs and economic development. Rhodesia had significant coal deposits, particularly in the Hwange and Wankie areas. Coal mining was also significant in Rhodesia. Coal mining was an important sector in Rhodesia, contributing to the country's energy needs and industrial development. Rhodesia had significant coal deposits, particularly in the Hwange and Wankie regions.

However, political and economic changes after Rhodesia gained independence from British colonial rule in 1980 and became Zimbabwe impacted the coal mining industry. The government introduced new policies and regulations, and the industry experienced shifts in ownership and management.

The coal mining industry in Zimbabwe, formerly Rhodesia, has continued to be an important sector in the country's energy mix. Hwange Colliery Company remains a significant coal producer, although the industry has faced various challenges, including infrastructure limitations, aging equipment, and financial constraints.

The Wankie Colliery Company, now known as Hwange Colliery Company, was the major player in coal mining in Rhodesia. It operated large-scale coal mines and played a crucial role in meeting the country's domestic coal demand and supporting various industries.

Coal mined in Rhodesia was used for multiple purposes, including electricity generation, industrial processes, and domestic heating. The coal industry provided employment opportunities and contributed to the growth of local communities around the mining areas.

The Wankie Colliery, located in the Wankie area (now known as Hwange), was the primary coal mining operation in Rhodesia. The coal extracted from the region was of high quality and suited for both domestic consumption and export.

Coal mining in Rhodesia played a vital role in meeting the country's energy requirements. Coal was used as a source of fuel for electricity generation, industrial processes, and domestic heating. The Wankie Colliery supplied coal to power stations and industrial facilities, contributing to the overall energy infrastructure of Rhodesia.

The coal mining industry also created employment opportunities and contributed to economic growth in the regions where mining activities were concentrated. Additionally, coal was exported to other countries, generating foreign exchange earnings for the country.

However, similar to other sectors in Rhodesia, the coal mining industry faced challenges over time. Factors such as changes in ownership, fluctuations in global coal prices, and shifts in energy policies influenced the industry's operations and investment climate. The country had coal deposits in areas such as Hwange, Wankie, and Wankie Colliery, which played a role in meeting domestic energy needs and supporting industrial activities. The country had coal deposits in the Hwange coalfield, located in the northwest region. The Hwange Colliery Company was the major coal producer and supplied coal for both domestic use and export.

These minerals, along with others such as nickel and copper, played a vital role in Rhodesia's economy and export earnings. The mining sector provided employment opportunities and contributed to the country's overall economic development. These mining activities, including gold, chrome, asbestos, and coal production, contributed to Rhodesia's economy, employment, and export revenues. However, it's important to note that the mining industry in Rhodesia faced challenges and changes over time due to factors such as political instability and economic shifts.

Steel manufacturing in Rhodesia

Rhodesia, now known as Zimbabwe, had a significant steel manufacturing industry during its existence as a self-governing colony and later as an unrecognized state. The steel industry played a vital role in the economic development of Rhodesia, contributing to its industrialization and infrastructure growth. Rhodesia, which is now known as Zimbabwe, had a significant steel manufacturing industry during the colonial era. Steel production played an important role in the country's economy and industrial development. Steel production did play a significant role in the country's economy and industrial development in Rhodesia.

Steel production played a crucial role in the economy and industrial development of Rhodesia. The establishment of steel mills and the production of steel products contributed to the growth of various sectors and supported the country's infrastructure development. The country's economy was primarily driven by agriculture, mining, and other sectors, with some emphasis on large-scale steel production.

Steel is a fundamental material used in construction, machinery, transportation, and other industries. In Rhodesia, the availability of locally produced steel helped meet the demand for infrastructure projects, including building bridges, roads, railways, and buildings. It also supported the development of the mining industry, as steel was needed for mining equipment and machinery. While Rhodesia did have some industrial activities, steel manufacturing was a growing part of its industrial landscape. The country also relied on importing steel and other manufactured goods for its industrial needs.

The steel industry in Rhodesia created employment opportunities and contributed to skill development. The operation of steel mills required a skilled workforce, which led to the training and employment of engineers, technicians, and other related professionals. The industry also provided indirect employment through the supply chain and support services.

Moreover, steel production contributed to the country's export earnings. Rhodesia could export steel products to neighbouring countries and beyond, generating revenue and improving its balance of trade.

The steel manufacturing sector played a vital role in driving industrialization, supporting economic growth, and fostering self-sufficiency in Rhodesia's development. However, as mentioned earlier, the political and economic challenges faced by the country after independence had a significant impact on the steel industry's growth and sustainability.

Rhodesia had a small but important steel manufacturing industry, primarily centred around the Zimbabwe Iron and Steel Company (ZISCO) located in the town of Redcliff. ZISCO was established in 1942 and became a major integrated steel producer in Africa.

The Rhodesian Iron and Steel Commission (RISCOM) was established in 1942 as a government agency in Rhodesia (now Zimbabwe). Its primary objective was to promote the development of the iron and steel industry in the country. RISCOM played a pivotal role in the establishment of the iron and steel company that later became known as the Zimbabwe Iron and Steel Company (ZISCO).

RISCOM conducted extensive studies to assess the viability of establishing an iron and steel industry in Rhodesia. This involved evaluating the availability of raw materials, infrastructure requirements, and market demand.

RISCOM was responsible for formulating development plans and strategies for the iron and steel industry. It played a key role in identifying suitable sites for steel plants, acquiring land, and coordinating infrastructure development for the industry. RISCOM sought financial support and investment for the establishment and growth of the iron and steel industry. This involved engaging with potential investors, both local and international, and facilitating the acquisition of funds for infrastructure development and plant construction.

RISCOM collaborated with international steel companies and experts to acquire technical knowledge and expertise. This involved engaging with consultants, engineers, and advisors to assist with the design, construction, and operation of iron and steel plants. RISCOM played a role in developing policies and regulations related to the iron and steel industry. It worked closely with government agencies to ensure supportive legislation and incentives for the growth of the sector.

The Rhodesian Iron and Steel Commission laid the foundation for the development of the iron and steel industry in Rhodesia, which later evolved into the Zimbabwe Iron and Steel Company (ZISCO). ZISCO became a significant player in steel manufacturing in the region and contributed to the country's industrial development.

The establishment of a steel manufacturing sector in Rhodesia can be traced back to the early 1940s. The Rhodesian Iron and Steel Commission (RISCO) was established in 1942 to explore the country's iron ore reserves and develop a steel industry. RISCO conducted extensive surveys and identified the Ripple Creek mine in the Midlands province as a potential source of iron ore.

In the mid-20th century, Rhodesia had a few steel mills that were established to meet the growing demand for steel products. The largest and most well-known steel mill was the Zimbabwe Iron and Steel Company (ZISCO), located in the town of Redcliff. ZISCO was established in 1942 and became one of the largest integrated steel producers in Africa.

In the 1950s, the Rhodesian government collaborated with private companies to establish the Rhodesian Iron and Steel Company (RISCO), which later became Zimbabwe Iron and Steel Company (ZISCO). ZISCO was officially formed in 1946 and constructed a steel plant near Redcliff, about 200 kilometres southwest of the capital, Salisbury (now Harare). The plant began production in 1947, initially focusing on the production of pig iron and later expanding into steel production. ZISCO operated blast furnaces, steelmaking plants, and rolling mills, producing a wide range of steel products, including billets, bars, rods, plates, and sheets. The company played a vital role in supplying steel for various sectors, such as construction, infrastructure development, agriculture, and manufacturing.

ZISCO played a crucial role in the industrial development of Rhodesia. The company supplied steel for various sectors, including construction, mining, agriculture, and infrastructure projects. It produced a range of steel products, including steel plates, sheets, rods, and pipes. ZISCO's steel products were used in the construction of railways, bridges, buildings, and other infrastructure projects across the country.

During its peak, ZISCO had an annual production capacity of around one million metric tons of steel. It employed thousands of workers and contributed significantly to Rhodesia's economy. The steel industry provided job opportunities, technical skills development, and contributed to the overall industrialization of the country. During its peak, ZISCO employed thousands of workers and contributed significantly to the national economy. However, political and economic challenges, including international sanctions and the escalation of the Rhodesian Bush War, affected the industry. The steel plant faced difficulties in accessing raw materials, spare parts, and international markets, leading to production disruptions and financial strain.

However, political and economic challenges in Rhodesia, particularly after its independence from British colonial rule in 1980, impacted the steel industry. Economic sanctions, political instability, mismanagement, and a decline in infrastructure development resulted in a gradual decline of the steel sector.

However, it's worth noting that after the country gained independence from British colonial rule in 1980 and became Zimbabwe, the steel industry faced various challenges. Economic and political factors, including sanctions, mismanagement, and a decline in infrastructure development, affected the industry's growth and sustainability.

Following the independence of Zimbabwe in 1980, ZISCO continued to operate under the ownership of the Zimbabwean government. However, the steel industry faced numerous challenges in the following decades, including mismanagement, lack of investment, and a decline in infrastructure maintenance. These factors, coupled with the broader economic issues faced by Zimbabwe, resulted in the decline of the steel industry.

In recent years, there have been efforts to revive and restructure the steel industry in Zimbabwe, including the ZISCO Steel Revival Project. However, the success of these initiatives remains uncertain, and the steel manufacturing sector in Zimbabwe continues to face significant challenges.

It's important to note that the information provided reflects the situation in the steel manufacturing sector. In recent years, the steel industry in Zimbabwe has faced various challenges, including aging infrastructure, inadequate investment, and limited access to raw materials and energy sources. ZISCO has struggled with financial difficulties and production constraints, leading to a significant reduction in its output. The current status and developments in the steel manufacturing industry in Zimbabwe, particularly in relation to Rhodesia, is constantly changing.

Rhodesia, which is now known as Zimbabwe, had a diverse economy that relied heavily on agricultural exports, particularly tobacco, maize, and cotton. Mining was also a significant sector, with notable resources such as gold, chrome, asbestos, and coal being extracted and exported.

While Rhodesia did have some limited steel production facilities, they were growing in scale and did contribute significantly to the country's industrial development. The economy also primarily revolved around the agricultural and mining sectors, which played more substantial roles in the country's economic land. The steel industry in Rhodesia contributed to various sectors of the economy, including construction, infrastructure development, agriculture, and manufacturing. Steel was used in the construction of buildings, bridges, railways, and other infrastructure projects. It also supplied steel products for the agricultural sector, such as farm equipment and machinery. Mining was also an important sector, with Rhodesia being a major producer of minerals like gold, chrome, asbestos, and coal.

The presence of a steel industry provided job opportunities and helped develop technical skills in Rhodesia. It played a vital role in the country's industrialization, as steel is a fundamental material for manufacturing and economic growth.

The future

Advances in extractive metallurgy have enabled the efficiency, sustainability, and economic viability of metal extraction processes of a wide range of metals, contributing to the development and advancement of numerous industries. Modern mining techniques have improved the efficiency and safety of ore extraction. Technologies such as remote sensing, GPS, and geological modelling help identify mineral deposits and optimize the location and design of mines. Automation and robotics have enhanced mining operations by increasing productivity and reducing the exposure of workers to hazardous environments.

Innovations in mineral processing have led to more effective methods of separating valuable minerals from the gangue. Advanced comminution techniques, such as high-pressure grinding rolls (HPGR) and stirred mills, provide energy-efficient methods of reducing particle size. Sophisticated mineral separation technologies, such as sensor-based sorting and advanced flotation systems, enable more precise and selective extraction of minerals.

Concerns about environmental impact and resource conservation have driven the development of sustainable processing technologies. These include hydrometallurgical processes, such as heap leaching and bioleaching, which minimize the use of hazardous chemicals and reduce water consumption. Recycling and reprocessing technologies have also been developed to recover valuable metals from secondary sources, such as electronic waste and scrap materials.

Electrowinning and electrorefining technologies have been refined and optimised for the extraction and purification of metals. These processes utilize electrolysis to selectively deposit metal ions onto electrodes or remove impurities from the metal. They have become essential in the production of high-purity metals, such as copper, zinc, and nickel, with stringent quality requirements.

Advances in computational modelling and simulation techniques have revolutionised process optimization and design in extractive metallurgy. Computer-aided process modelling allows for better understanding of the complex phenomena occurring during mineral processing and refining. It enables the prediction of process performance, optimisation of operating parameters, and reduction of costs and environmental impact through virtual experimentation.

The integration of renewable energy sources, such as solar and wind, into extractive metallurgy processes has gained attention. This helps reduce the carbon footprint and dependence on fossil fuels, making the operations more environmentally sustainable. Renewable energy can be utilized for power generation, heating, and other energy-intensive processes in the metal extraction and refining industry.

These advancements in extractive metallurgy have improved the efficiency of metal extraction, reduced environmental impact, and enhanced the sustainability of mining and metal production. Continued research and development in this field are crucial for further improvements in resource utilization, energy efficiency, and environmental stewardship in the mining and metallurgical industries.

The future of mining in Zimbabwe will depend on economic and political stability. Favourable policies, investment climate, and regulatory frameworks can attract both domestic and foreign investments in the mining sector. Political stability and government support for the industry are crucial for its growth and development.

The mining industry is continuously evolving, with advancements in technology leading to improved exploration techniques, efficient extraction methods, and enhanced environmental practices. Embracing new technologies can drive productivity, reduce costs, and improve sustainability.

The global prices of minerals and commodities play a significant role in shaping the mining industry's future. Demand and supply dynamics, geopolitical factors, and market trends can impact the viability and profitability of mining projects. The global demand for minerals and their prices will significantly impact the mining sector in Zimbabwe. If there is a sustained demand and favourable prices for key minerals found in the country, it can drive investment and exploration activities. Technological advancements can play a significant role in the future of mining. Innovations in mining techniques, equipment, and processes can improve efficiency, reduce environmental impact, and enhance the profitability of mining operations.

Increasingly, mining companies are expected to adhere to higher environmental and social sustainability standards. Environmental regulations, community engagement, and responsible mining practices will be important considerations for the future of mining in Zimbabwe.

A stable economic and political environment is crucial for attracting investments and fostering the growth of the mining sector. Policies that promote transparency, predictability, and a favourable business climate can positively impact the future of mining. Increasingly, environmental and social factors are being integrated into mining practices. Sustainable mining practices, community engagement, and environmental stewardship are becoming essential components of the industry. Compliance with regulations and adoption of responsible mining practices can influence the future of mining.

The government's policies and regulations related to the mining sector will shape its future. Clear and transparent regulations, fair licensing processes, and effective governance can contribute to a conducive environment for mining activities. Clear and consistent mining policies, regulatory frameworks, and legal systems are important for providing certainty and attracting investment. Governments that prioritize the mining sector and create an enabling environment can foster its future growth.

Adequate infrastructure, including transportation networks, power supply, and water resources, is critical for mining operations. Investments in infrastructure can unlock the potential of mineral resources and attract mining companies to a region.

It's worth noting that the future of mining in Zimbabwe, including its relationship to the country's history as Rhodesia, will be influenced by a range of factors that are subject to change and uncertainties. Favourable developments in the country's economic, political, and regulatory landscape will provide insights into the future direction of the mining industry. It's important to note that the future of mining in any region, including Rhodesia (Zimbabwe), depends on a complex interplay of these factors and can be influenced by various external and internal dynamics.

www.ingramcontent.com/pod-product-compliance
Lightning Source LLC
Chambersburg PA
CBHW080818220526
45466CB00011BB/3609